RAND NATIONAL DEFENSE RESEARCH INSTITUTE

A Wage Differential Approach to Managing Special and Incentive Pay

James Hosek, Michael G. Mattock, Beth J. Asch

Prepared for the Office of the Secretary of Defense

Approved for public release; distribution unlimited

For more information on this publication, visit www.rand.org/t/RR2101

Library of Congress Cataloging-in-Publication Data is available for this publication.
ISBN: 978-1-9774-0129-8

Published by the RAND Corporation, Santa Monica, Calif.
© Copyright 2019 RAND Corporation
RAND® is a registered trademark.

Cover: GettyImages / CatLane and Fanatic Studio.

Support RAND
Make a tax-deductible charitable contribution at
www.rand.org/giving/contribute

www.rand.org

Preface

The Office of Compensation within the Office of the Under Secretary of Defense for Personnel and Readiness asked RAND to conduct research into special and incentive (S&I) pays, which add to military cash compensation and serve a variety of force management purposes. Of particular interest was whether the existing structure of S&I pays might be improved by converting some pays into a "wage differential." Under this concept, S&I pays would be disbursed according to a stable schedule that could depend on occupation or duty, year of service, and pay grade. Two examples of a wage differential are Sea Pay and Aviation Career Incentive Pay (ACIP), both of which are paid according to schedules. A potential advantage of the wage differential approach is to provide greater stability in military compensation. The present research reviewed S&I pays to identify possible candidates for a wage differential, consider the cost-effectiveness of incentives embedded in certain S&I pays to select a longer period of obligated service (incentives absent under a wage differential), and estimate the value to the servicemember of decreasing S&I pay uncertainty.

This research should interest the policy community concerned with the design and effectiveness of military compensation, as well as the research community concerned with human resource and personnel issues.

The research was sponsored by the Office of the Under Secretary of Defense for Personnel and Readiness, conducted within the Forces and Resources Policy Center of the RAND National Defense Research Institute, a federally funded research and development center sponsored by the Office of the Secretary of Defense, the Joint Staff, the Unified Combatant Commands, the U.S. Navy, the U.S. Marine Corps, the defense agencies, and the defense intelligence community.

For more information on the RAND Forces and Resources Policy Center, see http://www.rand.org/nsrd/ndri/centers/frp.html or contact the director (contact information is provided on the web page).

Contents

Figures and Tables

Summary

Special and incentive (S&I) pays allow the Department of Defense (DoD) to address temporary personnel strength fluctuations, persistent differences between external pay and regular military compensation, personnel requirements for high retention in certain occupations, onerous and dangerous conditions such as imminent danger, and variations in external employment opportunities.

The Office of Compensation within the Office of the Under Secretary of Defense for Personnel and Readiness asked RAND to conduct research into the concept of a "wage differential." Under this concept, S&I pays would be converted into scheduled pay where the schedule would be stable over time and could depend on occupation, pay grade, and year of service. RAND research focused on (1) reviewing S&I pays to identify candidates for a wage differential; (2) providing examples of how a wage differential might be implemented in several occupations; (3) assessing the cost-effectiveness of incentives to select a longer obligation that are part of some S&I pays but would be absent from a wage differential; and (4) assessing the value to a servicemember of eliminating S&I pay uncertainty associated with reenlistment bonuses.

Reviewing S&I Pays to Identify Candidates for a Wage Differential

The military has over 50 types of active-duty special and incentive (S&I) pays. The Tenth Quadrennial Review of Military Compensation placed these pays into eight categories (U.S. Department of Defense, 2008, p. 46):

- health professions officers
- nuclear-qualified officers
- aviation-related officers
- officer accession bonus and retention incentives
- enlisted member enlistment, reenlistment, and retention bonuses
- hazardous duty
- assignment and special duty
- skill incentive or proficiency.

We reviewed the individual pays in these categories and found that they could be further classified with respect to four attributes: (1) whether the pay provides an occupational differential; (2) whether the pay contains an incentive to select a longer service obligation; (3) whether the pay compensates for hazardous duty or language proficiency; and (4) whether the pay is conditional on military circumstances.

> *Occupational differential S&I pays* are disbursed according to a schedule that depends on occupation, years of service, and pay grade. These include Sea Pay; Submarine Pay; and so-called incentive pays for health professions officers, nuclear officers, other officers, and enlisted aviators.

> *Pays containing an incentive to select a longer obligation* include so-called retention bonuses for health professions officers, nuclear-qualified officers, aviation-related officers, enlistment and reenlistment bonuses, as well as pays for warfare officers extending periods of active duty, judge advocate continuation pay, retention incentive for critical skills (officers), and the critical skills retention bonus (enlisted). Retention bonuses may vary in availability and amount depending on supply-and-demand conditions—for example, force growth or external employment conditions.

> *Hazardous duty pays* are for parachute duty, demolition duty, pressure chamber duty, and flight deck duty, among other duties. These pays are each a flat amount per month. There are also

hazardous duty pays for flying duty and air weapons controllers, which are paid according to a schedule that depends on duty, pay grade, and years of service. *Language proficiency pays* differ by language and require proficiency certification.

Pays that depend on military circumstances include Hostile Fire/ Imminent Danger Pay; pays for certain missions, locations, or tempos; Assignment Incentive Pays; Overseas Tour Extension Incentive Pays; pays for officers holding positions of unusual responsibility; Special Duty Assignment Pay for enlisted members; and incentive pays to change occupation, transfer between the Active Component and the Reserve Component within a service, and transfer between branches of service.

Occupational differential pays are good S&I pay candidates for a wage differential but are already virtual wage differentials. S&I pays that include an incentive to select a longer obligation are not good candidates because converting to a wage differential would remove this incentive and be less cost-effective. S&I pays for hazardous duty or proficiency typically come at a fixed rate per month—a simple "schedule." They do not depend on occupation but are akin to occupational differentials because they depend on specific criteria—namely, duty or proficiency. In this sense, they too are virtual wage differentials. S&I pays that vary over time due to economic conditions or force size fluctuations, such as enlistment and reenlistment bonuses, or that compensate for military circumstances that are ex ante uncertain, such as hostile deployment or certain assignments or locations, are not good candidates for a wage differential because they would be less cost-effective.

Examples of How a Wage Differential Might Be Implemented

One way to implement a wage differential would be to aggregate the S&I pays that are stable over time within selected occupations that have high average S&I pays and account for a large share of overall S&I

expenditures. By "stable over time" we mean those pays that remain largely constant in real terms for a given occupation, rank, and year of service and do not vary in response to change in requirements, the external labor market, work environment, or servicemember behavior. The wage differential would be constructed by summing the fixed component of S&I pays that are due to constant high external pay, burdensome but predictable duty, or high military training investment. S&I pays that involve incentives to select a longer obligation or that are conditional on military circumstances would be held apart.

To illustrate, the Navy offers a $15,000 accession bonus to officers selected for training in the Nuclear Propulsion Program, plus $2,000 upon successful completion of the program. After completing the initial service obligation, a nuclear-qualified, unrestricted line officer is eligible for a nuclear career annual incentive of $12,500. The example assumes that the officer receives this amount throughout a 26-year career. In addition, after completion of the initial service obligation the officer can sign up for nuclear-qualified officer continuation pay. The officer selects a service obligation of three, four, five, six, or seven years beyond the existing obligation and receives $35,000 per year, except for the first three-year contract, which pays $17,500 per year. We assume the initial contract is for three years. In addition to these pays, the officer is assigned to a submarine and receives Sea Pay and Submarine Pay.

As shown in Figure S.1, after the accession bonus is paid, the sum of S&I pays is fairly low in the first few years of service. It increases to over $40,000 in year of service (YOS) 5 and is in the range of $60,000 per year for YOS 8 through YOS 26. Figure S.2 shows how the wage differential concept might be implemented. The accession bonus, Sea Pay, Submarine Pay, and incentive bonus are aggregated to form the wage differential, with only nuclear officer incentive pay (labeled "Continuation" in the figure) outside the wage differential. The latter pay depends on a service obligation, assumed in the figure to be three years for the initial contract, and four or more years for the remainder of an officer's career. In the example, the wage differential portion of S&I pay is fairly stable at $25,000 to $30,000 between YOS 5 and YOS 25.

Figure S.1
Notional Example of Nuclear-Qualified Submarine Officer S&I Pay

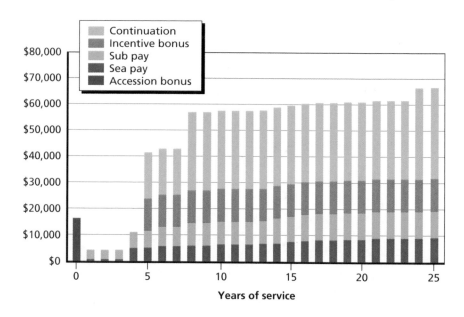

Figure S.2
Notional Example of Nuclear-Qualified Submarine Officer S&I Pay
Under a Wage Differential Approach

Cost-Effectiveness of Incentives to Select a Longer Obligation

One possible concern with setting a substantial wage differential as a fixed, unconditional part of compensation is that if it included retention bonus pays, it would eliminate the incentive in them to select a longer obligation. We explore this issue by using RAND's Dynamic Retention Model of U.S. Air Force (USAF) pilot retention behavior to simulate policy alternatives (Mattock et al., 2016). The primary S&I pays for this occupation come in two forms: an occupational differential and a retention bonus, the amount of which depends on service commitment. Air Force pilots are eligible for Aviation Career Incentive Pay (ACIP), an unconditional monthly pay of up to $840, and an aviator retention bonus (also called Aviator Continuation Pay, ACP), which is conditional on committing to a multiyear contract. In our example, we assume ACP paid up to $25,000 per year of obligated service. (In 2016, ACP was increased to $35,000 per year of obligated service.)[1]

Under the first scenario, ACP is eliminated and ACIP is increased to restore pilot end strength to the baseline level. Under the second scenario, ACIP is eliminated and ACP is increased to compensate. The purpose of considering these extreme scenarios—no ACP versus no ACIP—is not because these are realistic scenarios or ones that we would recommend. Rather, they show the upper and lower bounds for setting the wage differential, the additional cost of rolling into the wage differential a part of S&I pay that currently depends on a contract obligation, and the savings of reducing the wage differential and rolling part of the wage differential into a part that depends on an obligation.

We find that ACIP, the wage differential portion of S&I pay, would have to increase by 264 percent to sustain retention relative to

---

[1] In FY 2017, ACIP was renamed Aviation Incentive Pay or AvIP and ACP was renamed Aviation Bonus or AvB (http://www.esd.whs.mil/Portals/54/Documents/DD/issuances /dodi/773067_dodi_2016.pdf). Furthermore, the maximum monthly amount of ACIP was increased to $1,000. Throughout this report, we use the former names ACP and ACIP and the $840 monthly maximum.

Table S.1
Steady-State ACIP and ACP Costs, 2014 Dollars

	Percentage Change in ACIP to Sustain Retention	Percentage Change in ACP to Sustain Retention	Baseline Cost	Policy Scenario Cost	Difference	Percentage Difference
Scenario 1 (All ACIP)	264%	NA	$64,203,370	$170,879,300	$106,675,930	166.2%
Scenario 2 (All ACP)	NA	27%	$64,203,370	$33,184,650	–$31,018,720	–48.3%

the baseline in the steady state.[2] On the other hand, we find that ACP would have to increase by 27 percent to sustain retention.

Table S.1 shows that baseline ACP and ACIP costs for the USAF rated community are $64.2 million in 2014 dollars. Under scenario 1, these costs increase to $170.9 million, or by 166 percent—that is, more than doubling the cost. The reason for the cost increase is that eliminating ACP means that S&I pay no longer depends in part on a service obligation, so the retention effect of the S&I pay decreases. ACIP would have to increase by more than a dollar-for-dollar rate when ACP is eliminated because ACIP does not entail a service obligation. More generally, the results imply that a blend of S&I pay that favors a wage differential and rolls into that wage differential all or part of S&I pay that depends on a contract obligation will increase costs. The table also shows the other extreme case: eliminating ACIP. In this case, costs decrease to $33.2 million, or by 48 percent, implying that a blend of S&I pay that favors making S&I pay contingent on a contract obligation is more cost-effective.

---

[2] By "steady state," we mean a case in which all servicemembers have spent their entire careers under the policy environment being considered.

Value to a Servicemember of Eliminating Reenlistment Bonus Uncertainty

We use an expected utility model to show that when an individual is risk averse and pay is uncertain, it is more cost-effective to compensate conditionally on the realization of the uncertain outcome rather than to pay a fixed amount up front. In a military context, it is more cost-effective to compensate for military circumstances such as hostile deployment or certain assignments or locations, or for variation in supply-and-demand conditions driven by change in force size or the unemployment rate, when these circumstances and conditions arise rather than by a fixed schedule such as a wage differential. That is, it is not cost-effective to compensate for uncertain circumstances in the form of a wage differential.

We also estimate the value to a servicemember of eliminating uncertainty in reenlistment bonuses. Risk aversion is modeled by a constant relative risk aversion utility function, and we base the value of the risk parameter on an average of estimates derived from published studies. We draw on earlier RAND research on military cash compensation for values of the standard deviation of the reenlistment bonus. We find that the value to an entering servicemember of eliminating bonus uncertainty would be less than 1 percent of the present discounted value of expected pay over the first ten years of service.

Conclusion

This research examined the idea of converting S&I pays into a wage differential. The research classified S&I pays with respect to four attributes that were related to the suitability of using a wage differential and considered the role of incentives to select a longer obligation and the value to the servicemember of decreasing S&I pay uncertainty. The key findings are as follows.

First, a number of S&I pays (e.g., occupational pays and hazardous duty pays) already have features like a wage differential. Second, incentives to select a longer obligation contribute to the cost-effectiveness of

S&I pay, and eliminating these incentives—as would be done under a wage differential—would decrease cost-effectiveness. Third, in the presence of risk aversion, it is cost-effective to compensate for uncertain circumstances such as hostile deployment and variation in supply-and-demand conditions by paying S&I pay when the circumstances are realized rather than on a continuous, scheduled basis, as would be done under a wage differential. Fourth, even if it were thought desirable to convert such pays into a wage differential, the value to the servicemember of doing so would be small.

Overall, the research provides new insight into the structure and cost-effectiveness of existing S&I pays. The findings indicate that a wage differential is already present in many S&I pays and that other pays, which are paid conditional on circumstances, are more cost-effective in their current form than if they were paid as a wage differential. Finally, the findings indicate greater cost-effectiveness when S&I pay includes an incentive to select a longer obligation, which therefore suggests going in the direction of making greater use of such incentives rather than in the direction of a wage differential.

Acknowledgments

We are grateful to Jeri Busch, Director, Office of Compensation within the Office of the Under Secretary of Defense for Personnel and Readiness, and Don Svendsen of that office, for their guidance and support throughout this project. We are pleased to thank our RAND colleagues Whitney Dudley, Chris Guo, Anthony Lawrence, Brian Stucky, and Cate Yoon, who contributed to the project at various stages. We also thank Alex Rothenberg of the RAND Corporation and Paul Hogan of the Lewin Group for their thorough, helpful reviews of this work.

Abbreviations

AC	Active Component
ACCP	Aviator Career Continuation Pay
ACIP	Aviation Career Incentive Pay
ACP	Aviator Continuation Pay
ACRB	Aviator Command Retention Bonus
ADSC	active duty service commitment
ARP	Aviator Retention Pay
CEFIP	Career Enlisted Flyer Incentive Pay
CRRA	constant relative risk aversion
CSP	Career Sea Pay
DFAS	Defense Finance and Accounting Service
DMDC	Defense Manpower Data Center
DoD	Department of Defense
DRM	dynamic retention model
FY	fiscal year
HPSP	Health Professions Scholarship Program
NOBC	Navy Officer Billet Classification
QRMC	Quadrennial Review of Military Compensation
RC	Reserve Component
ROTC	Reserve Officers' Training Corps

S&I	special and incentive
USAF	U.S. Air Force
YOS	year of service

CHAPTER ONE

Introduction

This research explores the idea of introducing a wage differential into the system of special and incentive (S&I) pays. A wage differential would provide supplemental compensation according to a schedule that could depend on occupation or duty, pay grade, and years of service. The wage differential would replace some of today's S&I pays, while S&I pays not included in the wage differential could be allocated as they are today. The wage differential could be achieved through separate basic pay tables that vary by occupation or duty. However, our analysis considers only a differential that is additive to a servicemember's compensation and is not necessarily tied to basic pay.

S&I pays are a relatively small but important portion of military cash compensation, which consists of regular military compensation, S&I pay, and certain allowances (e.g., the family separation allowance, uniform allowance, or overseas housing allowance). Regular military compensation includes basic pay, basic allowance for subsistence, basic allowance for housing, and an adjustment deriving from the nontaxability of the allowance. In 2017 cash compensation totaled $25.8 billion for officers and $58.8 billion for enlisted personnel, and S&I pays were respectively 6.1 percent and 4.4 percent of the total (see Table 1.1). In addition to cash compensation, the armed services contribute to the retirement system and pay the Social Security tax and separation pay.

A potential gain from a wage differential would be to decrease S&I pay uncertainty related to the availability and amount of the pay. The present research looks at the following specific questions related to the wage differential concept:

Table 1.1
Military Cash Compensation, FY 2017 (in billions of dollars)

Type	Officers	Enlisted
Basic pay	17.456	35.296
Basic allowance for housing	5.622	13.872
Basic allowance for subsistence	0.702	4.922
Incentive pays	0.478	0.242
Special pays	1.095	2.324
Allowances	0.455	2.103
Total	25.808	58.760
S&I as percentage of total	6.1	4.4

SOURCE: U.S. Department of Defense, 2017.

- What are the existing S&I pays, and which of them would be good candidates for a wage differential?
- What are examples of how a wage differential might be implemented?
- Should S&I pay that includes an incentive to select a longer obligation be included in a wage differential?
- Should S&I pay that is conditional on military circumstances or supply-and-demand conditions be included in a wage differential?
- Related to that question, if such pays were included in a wage differential, what would the gain be to the servicemember?

We address these questions in the following chapters. Chapter Two presents attributes that are relevant to determining whether an S&I pay is a suitable candidate for a wage differential and classifies S&I pays by the attributes. Chapter Three offers examples of how S&I pays in selected occupations could be converted into a wage differential. Chapter Four considers the role of incentives to select a longer obligation with respect to the cost-effectiveness of S&I pay. We use RAND's dynamic retention model (DRM) to make a quantitative estimate of the extent to which the incentives contribute to cost-effectiveness. The first part of Chapter Five presents an argument for why it is more cost-effective to disburse S&I pays that are conditional on uncertain circumstances when the circumstances are realized rather than through an unconditional pay schedule such as a wage differential; the second part provides an estimate of the gain to the servicemember if the uncertainty of these pays were eliminated, as would be the case if they were paid via a wage differential. The final chapter, Chapter Six, offers our conclusion.

CHAPTER TWO

Which Special and Incentive Pays Are Suitable for a Wage Differential?

Military assignments differ in their nature of work, work conditions, and risk of danger. The military uses 50 types of active-duty S&I pay to compensate for these differences. To consider which S&I pays are suitable for the wage differential concept, it is useful to classify them in broader categories.

The Tenth Quadrennial Review of Military Compensation (QRMC) proposed eight categories (U.S. Department of Defense, 2008, p. 46):

- health professions officers
- nuclear-qualified officers
- aviation-related officers
- officer accession bonus and retention incentives
- enlisted member enlistment, reenlistment, and retention bonuses
- hazardous duty
- assignment and special duty
- skill incentive or proficiency.

When the Tenth QRMC issued its report in 2008, there were over 60 S&I pays. Many of these were for health professions officers, and they have recently been consolidated into two pays, incentive pay and a retention bonus. Still, the eight categories remain a useful classification scheme. The Tenth QRMC recommended grouping S&I pays into the eight categories as a way of giving the armed services more flexibility in allocating the total dollars in each category to S&I pays in the category: "Within each category, the Services would have flexibility to allocate resources to those areas that would most effectively and efficiently meet staffing needs" (U.S. Department of Defense, 2008). The individual pays would continue to have their own identity, purpose, eligibility requirements, and pay schedules. The Tenth QRMC further argued that the S&I pay budget (in 2008) might be too small to meet force management requirements in the future. It therefore recommended increasing the overall S&I pay budget and proposed allocating to S&I pays "the portion of future pay raises that exceeds the employment cost index," subject to the proviso that comparability between military and civilian pay existed (U.S. Department of Defense, 2008). It seems clear that the Tenth QRMC recognized the potency and flexibility of S&I pays as a force management tool, despite their relatively small fraction of the compensation budget.

Like the Tenth QRMC, the Eleventh QRMC recognized the role of S&I pay in meeting military manpower requirements. Rather than offering categories to group S&I pays, the Eleventh QRMC offered five rationales for S&I pay (U.S. Department of Defense, 2012, pp. 33–34):

- high civilian wages for similar skills
- rapid growth in demand
- onerous or dangerous working conditions
- high training investment costs
- special skills and proficiency.

An S&I pay can have one or more of these rationales. For example, the S&I pays for health professionals, nuclear engineers, and aviators compensate for high civilian wages and significant training investment costs paid by the military. Enlisted and officer management S&I pays mainly consist of accession and retention bonuses to meet manning demands when the demand for personnel increases, conditions in the military worsen, or external job opportunities improve. Pays for special skills and proficiency provide incentives to maintain human capital (as verified by recurrent certification) and sustain retention. Pays for hazardous duty pay, assignment, and special duty address burdensome, difficult, or dangerous working conditions and also help to retain personnel when there has been an investment in training for hazardous duties.

To identify S&I pays suitable for the wage differential concept, we propose four attributes to describe S&I pays and then cross-classify S&I pays using the Tenth QRMC's eight categories and these four attributes. The attributes are (1) whether the pay provides an occupational differential; (2) whether the pay contains an incentive to select a longer service obligation; (3) whether the pay compensates for hazardous duty or language proficiency; and (4) whether the pay is conditional on military circumstance.

An S&I pay has the *occupational differential* attribute when it is disbursed according to a schedule that depends on occupation, pay grade, and years of service. Sometimes these pays are called incentive pays (e.g., health professions officer incentive pay). Schedules are appropriate when there are persistent differences between external and military pay levels or persistent arduous or demanding conditions of duty.[1] For instance, Sea Pay and Submarine Pay compensate for the demands of duty at sea. (Submarine Pay may be disbursed in addition to Sea Pay.) These pays do not include an incentive to select a longer obligation.

Pays that include *an incentive for a longer obligation* are often referred to as accession and retention bonuses, and such pays are available for enlisted personnel through enlistment and reenlistment bonuses and for officers in certain occupations (e.g., through aviation

---

[1]   Wage differential schedules would need to take into account the active duty service commitment (ADSC) typically incurred when a service provides or funds training or education that is highly valued in the external labor market. For example, Undergraduate Pilot Training in the Air Force currently entails an ADSC of ten years. During the ADSC the servicemember is obligated to serve and typically would not receive pay that reflects persistent differences in the military wage and the external market wage until his or her service commitment is concluded. The underlying logic is that the servicemember has already received compensation in kind for service in the form of the valued training or education, and that he or she freely chose to be obligated in exchange for receiving the training or education. Thus, during the ADSC any occupational differential due to persistent differences between the military and civilian wage would generally be omitted from the wage differential paid to the servicemember.

retention bonuses or nuclear career incentive continuation pays). These pays offer more money for a longer obligation, and in some cases the pay rate (dollars per year) itself increases as the length of obligation increases. To receive these pays, the individual is obligated to some period of service—for instance, three or four years with a reenlistment bonus, or possibly nine years with an aviation retention bonus, depending on what is on offer.

Hazardous duty and language proficiency pays compensate for duties that involve hazardous conditions or require proficiency with a foreign language. The pays also help to sustain retention, which is valuable to the military given the training investment needed to qualify for such duty.

Pays that are *conditional on military circumstances* recognize that, regardless of occupation or duty, certain circumstances arise that are unusually dangerous or difficult. Pays corresponding to these circumstances include Hostile Fire/Imminent Danger Pay; pays for mission, location, or tempo; and Assignment Incentive Pay, among others.

Pays with an incentive for a longer obligation and pays conditional on military circumstances share a similarity: uncertainty. We could have included uncertainty as another attribute, but it seemed sufficient to note that the military circumstances in the *conditional on military circumstances* attribute are all uncertain ex ante, and the *incentive for a longer obligation* attribute can have uncertainty attached to it. Military circumstances such as the incidence and duration of deployments are uncertain, for instance. An incentive for a longer obligation can be present when there is little uncertainty, as in the case of external versus military pay for health professions officers, but in other cases supply-and-demand conditions can fluctuate and drive changes in the amount of pay offered, given that it comes with a service obligation. An improvement in civilian employment opportunities can shift back the supply of recruits and decrease the willingness to reenlist, for instance, while a directive to increase force size will increase the demand for personnel. In addition, uncertainty may cause interaction between pays. For instance, when Army deployments to Afghanistan and Iraq proved longer than expected, recruiting and retention began to decrease despite the disbursement of Hostile Fire Pay. Hostile Fire

Pay had a preset rate, so to further augment pay the Army increased bonus amounts and offered bonuses in more occupations.

Table 2.1 shows the cross-classification of S&I pays by the Tenth QRMC's eight categories and the four attributes. Cell entries in the table listing the S&I pays and selected features are based on the Department of Defense's authoritative guide (Under Secretary of Defense, 2017), which describes the types, eligibility conditions, amounts, and details of S&I pays. Inspection of the cross-classification shows that each S&I pay is associated with a single attribute. This, of course, is a reflection of the way we defined the attributes, though it also indicates that the pays can be mapped to the attributes and, in effect, place the S&I pays into four separate, nonoverlapping groups. This facilitates discussing the large number of S&I pays by attribute rather than individually—that is, this particular decomposition of S&I pays is useful for thinking about which pays are good candidates for incorporation into a wage differential.

Special and Incentive Pay Attributes and Suitability for a Wage Differential

The earlier discussion describing the attributes in some sense anticipates whether an S&I pay would be suitable for a wage differential.

S&I pays with the occupational differential attribute are good candidates for a wage differential. These pays exist because of high external pay, burdensome but predictable duty, or high military training investment. They are adjusted for comparability to market rates, and stability is thus in the form of comparability rather than a fixed amount. For health professions officers, nuclear-qualified officers, and aviation-related officers the wage differential is paid as incentive pay. For health professions officers it is a flat rate per year that varies by specialty. For nuclear officers it is $35,000 per year for an obligation of three to seven years. For aviation-related officers it is specified with respect to years of aviation service and increases from $125 per month to a peak of $840 per month, which is reached after 14 years of service and declines to $585 per month over 22 years of service and $250

Table 2.1
A Classification of Special and Incentive Pays

Category	Type	Pay Rate or Schedule	Occupational Differential	Incentive for Longer Obligation	Hazardous Duty or Language Proficiency	Conditional on Military Circumstances
Health professions officers	Incentive pay; varies by specialty	Flat rate per year	X			
	Accession bonus; varies by specialty	Up to $60,000		X		
	Critical Wartime Skills Accession Bonus; varies by specialty	$150,000 to $400,000; requires 4-year obligation		X		
	Retention bonus; varies by specialty	Higher rate/year for longer obligation		X		
Nuclear-qualified officers	Nuclear Career Accession Bonus	Up to $50,000 for each 12-month obligation		X		
	Nuclear Career Incentive Pay	Up to $35,000/year for 3-, 4-, 5-, 6-, or 7-year obligation	X	X		
	Nuclear officers extending period of active duty	Up to $30,000 for each year of obligation		X		
Aviation-related officers	ACIP	Schedule by years of aviation service; $125/month to $840/month	X			
	Remotely Piloted Aircraft Aviation Incentive Pay	Same schedule as for ACIP	X			
	Aviator Retention Bonus	Up to $35,000 per 12-month obligation		X		

Table 2.1—Continued

Category	Type	Pay Rate or Schedule	Occupational Differential	Incentive for Longer Obligation	Hazardous Duty or Language Proficiency	Conditional on Military Circumstances
Officer accession bonus and retention incentives	Career Sea Pay (CSP) and CSP Premium	Schedules for enlisted personnel, warrant officers, and officers	X			
	Submarine Duty Incentive Pay	Schedules for enlisted personnel, warrant officers, and officers	X			
	Accession bonus for new officers in critical skills	Up to $60,000	X			
	Accession bonus for officer candidates	Up to $8,000		X		
	Warfare officers extending period of active duty	Up to $15,000 for each year obligated		X		
	Surface Warfare Officer Continuation Pay	Up to $50,000 to remain on active duty and complete one or more tours of duty		X		
	Judge Advocate Continuation Pay	Career limit of $60,000; "for a period of obligated service"		X		
	Retention incentive for critical military skills	Career limit of $200,000 except for health professions officers		X		

Table 2.1—Continued

Category	Type	Pay Rate or Schedule	Occupational Differential	Incentive for Longer Obligation	Hazardous Duty or Language Proficiency	Conditional on Military Circumstances
Enlisted member enlistment, reenlistment, and retention bonuses	CSP and CSP Premium	Schedules for enlisted members, warrant officers, and officers	▓			
	Submarine Duty Incentive Pay	Schedules for enlisted members, warrant officers, and officers	▓			
	Career Enlisted Flyer Incentive Pay (CEFIP)	Schedule by years of aviation service, up to $400/month	▓			
	Remotely Piloted Aircraft Career Enlisted Aviation Incentive Pay	Same schedule as for CEFIP	▓			
	Diving Duty Pay	Up to $240/month for officers; $340/month for enlisted personnel	▓			
	Enlistment bonus	Up to $40,000		▓		
	Selective Reenlistment Bonus	Up to $25,000/year of obligated service; $100,000 maximum		▓		
	Critical Skills Retention Bonus	Up to $30,000; career limit of $200,000		▓		
Skill incentive or proficiency	Foreign Language Proficiency Pay	Up to $12,000 per 1-year certification			▓	

Table 2.1—Continued

Category	Type	Pay Rate or Schedule	Occupational Differential	Incentive for Longer Obligation	Hazardous Duty or Language Proficiency	Conditional on Military Circumstances
Hazardous duty	Flying duty, crew members	$150/month to $250/month; varies by pay grade			■	
	Hazardous Duty Incentive Pay for air weapons controller crew members	Schedule by grade/year of service (YOS); $150/month to $350/month			■	
	Flying duty, noncrew members	$150/month; $250/month for high altitude/low opening jumping			■	
	Parachute Duty Pay, Demolition Duty Pay, Pressure Chamber Duty Pay, Acceleration and Deceleration Duty Pay, Thermal Stress Duty Pay, Flight Deck Duty Pay, Toxic Pesticides/Dangerous Organisms Personal Exposure Pay, Toxic Fuel/Propellants and Chemical Munitions Exposure Duty Pay for visit, board, search, and seizure—maritime interdiction operations	$150/month; if qualified for more than one of these pays, may receive maximum of two			■	
	Hostile Fire/Imminent Danger Pay	$225/month				■
	Mission	$150/month				■
	Location	$50/month to $150/month depending on location				■
	Tempo	$450/month for sailors and marines deployed beyond the 220th day of a consecutive operational deployment				■

Table 2.1—Continued

Category	Type	Pay Rate or Schedule	Occupational Differential	Incentive for Longer Obligation	Hazardous Duty or Language Proficiency	Conditional on Military Circumstances
Assignment and special duty	Assignment Incentive Pay	Variable; $3,000/month cap				
	Overseas Tour Extension Incentive Pay	$80/month or $2,000 for a 12-month extension				
	Officers holding positions of unusual responsibility	$50/month, O-4 and below; $100/month, O-5; $150/month, O-6				
	Special Duty Assignment Pay for enlisted members	$75/month to $450/month depending on difficulty of assignment or unusual degree of military skill needed				
	Incentive bonus for conversion to military occupational specialty to ease personnel shortage	Up to $4,000				
	Incentive bonus for officers to transfer between armed forces	Up to $10,000				
	Incentive bonus for transfer between components	Up to $10,000				

SOURCE: Under Secretary of Defense, 2017; U.S. Department of Defense, 2008.

per month over 25 years of service. Sea Pay and Submarine Pay are disbursed by schedules based on pay grade and years of service. For example, Submarine Pay increases by pay grade and year of service for enlisted personnel, increases by year of service for the first few years for warrant officers and then stays at $425 per month for five or more years of service; it is independent of pay grade, and increases by year of service for officers (under O-7) and first increases then decreases by grade given years of service. Overall, these pays are stable, scheduled additions to regular military compensation. With this structure and stability, they are excellent candidates for a wage differential, but they are virtual wage differentials already.

S&I pays with an incentive to select a longer obligation are offered alongside the occupational differentials for health professions officers, nuclear officers, and aviation-related officers. The incentive is present both in accession bonuses and retention bonuses, and the bonus rates are stable over time, though they are periodically reset. Health professions officer retention bonuses vary by specialty and are unusual in that they offer a higher rate per year for longer obligations; the rate is specified in the Department of Defense (DoD) regulation. Nuclear and aviation-related officer retention bonuses have a cap that is specified in the regulation. For instance, retention bonuses for nuclear-qualified officers are capped at $30,000 per year of obligation, and aviation retention bonuses are capped at $35,000 per year of obligation. Other well-known types of bonuses are enlistment bonuses, reenlistment bonuses, and critical skills retention bonuses; these, however, are not accompanied by an occupational differential S&I pay, with the exception of personnel eligible to receive Sea Pay and Submarine Pay. The enlistment bonus is capped at $40,000, the reenlistment bonus is capped at $25,000 per year of obligation and has a maximum of $100,000, and the critical skills retention bonus is capped at $30,000 for an obligation of at least one year.

S&I pays with an incentive to select a longer obligation are the primary means by which the armed services can affect enlistment and retention behavior. The caps on the bonuses regulate the maximum generosity of the offer, and the services have flexibility to offer amounts up to the cap when enlistment and retention needs to be shored up or increased. By the same token, the bonuses can be decreased when they are not needed. Although we do not pursue this, it is likely that bonus rates vary little for health professions, nuclear, and aviation-related officers because of the stability of military/civilian wage differences in these occupational areas but vary more for enlisted personnel because their job opportunities are more affected by civilian employment conditions. This is apparent in Figure 2.1, which shows low unemployment and less increase in unemployment during the Great Recession for professional occupations than other occupations.

The role of the incentive for a multiyear commitment weakens if more of the pay is disbursed as a wage differential without condition. The analytical issue is how strong the incentive effect is. That is, to what extent is retention adversely affected as the retention incentive portion is converted into a wage differential without obligation? We address this question in Chapter Four with the DRM by comparing the cost of attaining a retention profile through S&I pay requiring a multiyear commitment versus S&I pay not requiring a multiyear commitment.

Figure 2.1
Unemployment Rate for Selected Occupations, 2000–2017

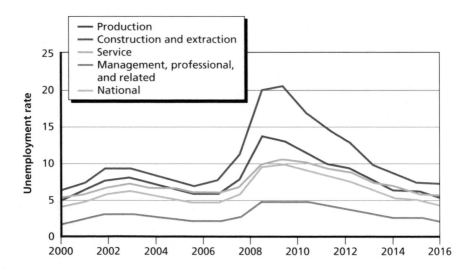

Hazardous duty and language proficiency pays are grouped under the same attribute. Hazardous Duty Incentive Pay comes at a fixed rate of $150 per month in most cases (e.g., parachute duty, demolition duty, and flight deck duty), though for air weapons controllers it is scheduled by year of service and pay grade. Foreign Language Proficiency Pay is up to $12,000 per one-year certification; more difficult languages receive higher pay. These pays are not related to external wages or changes in supply and demand but are stable rates and recognize the skill and knowledge required. The pays can be expected to have a positive effect on retention and on the decision to enter occupations requiring a certain skill or proficiency. Like occupational differentials, they are already similar to a wage differential, yet they are conditional on being assigned to a qualifying duty or maintaining language proficiency.

The final attribute relates to pays that are conditional on military circumstances. These pays include Hostile Fire/Imminent Danger Pay; Assignment Incentive Pay; pays for mission, location, and tempo; and pay for positions of unusual responsibility. These pays come at a fixed rate per month. There are also pays for transferring to other occupations, components, and branches of service, which are one-time pays of up to $4,000, $10,000, and $10,000, respectively. In Chapter Five we provide reasons why pays with this attribute are not suited to a wage differential.

To summarize, we have classified pays according to four attributes and used the attributes to discuss which pays might be good candidates for a wage differential. Pays that have the *occupational differential* and *hazardous duty or language proficiency* attributes appear to be good candidates, yet as mentioned, these pays are already virtual wage differentials. Pays with the *incentive for a longer obligation* attribute may or may not be good candidates, depending on the extent to which the incentive makes the pay more cost-effective (see Chapter Four). Additionally, these pays and pays with the *conditional on military circumstances* attribute may not be good candidates because they address circumstances that are uncertain (see Chapter Five).

CHAPTER THREE

Examples of Implementing a Wage Differential

Chapter Two argued that S&I pays suited to a wage differential concept are those that are stable over time, scheduled, and not involving an incentive to select a longer obligation or payment conditional on the realization of certain military circumstances. The chapter also argued that the best candidates for a wage differential are occupations with high S&I pay and that account for a large share of overall S&I pay expenditures.[1]

This chapter considers how the concept of a wage differential might be implemented in three occupations: a physician internist, a nuclear-qualified officer, and an aviator. The examples are for officers in occupations where existing S&I pays provide a predictable, persistent addition to regular military compensation. For each occupation, we illustrate S&I pay currently receivable in that occupation, then put some of these pays into a wage differential.

Internal Medicine

We assume the internist has a four-year health professions scholarship program (HPSP) obligation, the fulfillment of which begins after completing a three-year internship/residency in the military. The example assumes the residency is done in the first three years of military ser-

---

[1] Appendix C provides information on the extent to which S&I pay expenditures are concentrated among a small number of occupations versus being spread evenly across occupations.

vice, entailing an obligation of three years, though the residency and HPSP obligations can be fulfilled concurrently.[2] Thus, the four-year HPSP obligation dictates the extent of the obligation. This obligation is fulfilled after the seventh year of service has been completed. The example shows S&I pay through 26 years of service, though many physicians leave the military after their initial obligation.

Effective January 28, 2018, DoD consolidated many individual special pays, and a health professions officer may receive incentive pay and a retention bonus for one specialty.[3] Incentive pay is initially $8,000 per year for a not-yet-licensed internist and $43,000 per year for a licensed internist. We assume the internist becomes licensed in the second year of residency and begins to receive $43,000 in the third year. After completing the active duty service obligation, the internist continues to receive incentive pay and is now eligible for a retention bonus. The internist selects a retention bonus with a four-year obligation that pays $35,000 per year, which in our example starts in YOS 8. If instead the internist had chosen a two-year obligation, the annual rate would be $13,000; and if a three-year obligation were chosen, the annual rate would be $23,000. We assume the internist continues to select a four-year retention bonus as long as eligible, which requires being below the grade of O-7. Finally, we assume the internist becomes board certified in YOS 11 and can receive $6,000 per year in board certification pay that year. Figure 3.1 shows the internist's S&I pays by year of service.

There is no reason to create another figure to show a wage differential as it would look the same, only with incentive pay relabeled as wage differential (as was mentioned in Chapter Two, incentive pay functions as a wage differential). However, before the consolidation of

---

[2]   Hosek, Nataraj, Mattock, and Asch, 2017.

[3]   Before 2018 the special pays included Variable Special Pay in training (residency) and after completing training; Additional Special Pay, assuming the internist became licensed (e.g., in the second year of residency); Incentive Special Pay, receivable upon completing the residency; Early Career Incentive Special Pay, disbursable when the licensed internist was within 18 months of completing the initial obligation; Multiyear Special Pay, disbursable after completing the initial obligation; and Board Certification Pay if the internist became board certified, which might be in the tenth or eleventh year of service.

Figure 3.1
Notional Example of Internist S&I Pay over a 26-Year Career

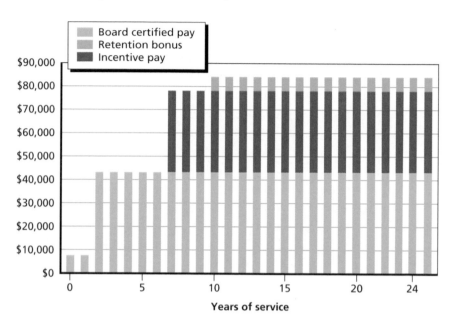

health professions officer S&I pays, the wage differential would have replaced variable special pay, the previous version of incentive special pay, and early career incentive special pay. In this case, then, events have overtaken the wage differential concept.

Nuclear-Qualified Personnel

The Nuclear Career Accession Bonus can be up to $50,000 per year of obligation. Today, the Navy offers a $15,000 accession bonus to officers selected for training in the Nuclear Propulsion Program, plus another $2,000 after successful completion of the program. Officers typically enter with a military service obligation. For instance, the obligation is five years for a service academy graduate, four years for a Reserve Officers' Training Corps (ROTC) scholarship recipient, and three years for a nonscholarship ROTC student. After completion of the initial obligation, a nuclear-qualified, unrestricted line officer is eligible for a nuclear career annual incentive of $12,500 (Under Secretary of Defense, 2017). The example presumes that the officer receives this amount throughout the assumed 26-year career, but the bonus can be paid at its maximum value of $22,000 per year for an O-6 with more than 26 years of service, or if the officer has served or is serving as an unrestricted line officer for a major command or is a major program acquisition professional. Additionally, after completion of the initial service obligation, the officer can sign up for Nuclear-Qualified Officer Continuation Pay. The officer selects a service obligation of three, four, five, six, or seven years beyond the existing obligation, and receives $35,000 per year except for the first three-year contract, which pays $17,500 per year. We assume the initial contract is for three years. In addition to these pays, the officer is assigned to a submarine and receives Sea Pay and Submarine Pay.

The nuclear-qualified officer's S&I pays in this example are shown in Figure 3.2. The sum of the pays is fairly low in the first few

Figure 3.2
Notional Example of Nuclear-Qualified Officer S&I Pay

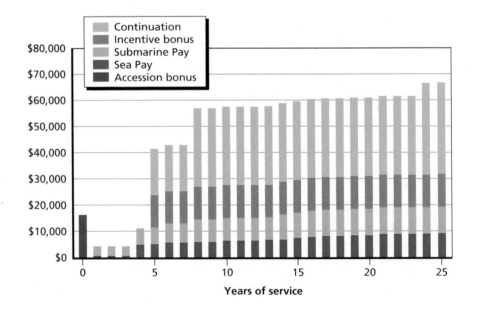

years of service but increases to over $40,000 in YOS 5 and is in the range of $60,000 per year for YOS 8 through YOS 26.

Figure 3.3 shows an example of how the wage differential concept might be implemented for nuclear-qualified officers. The accession bonus, Sea Pay, Submarine Pay, and incentive bonus are put in the wage differential, with only the nuclear officer incentive pay (labeled "Continuation" in the figure) outside the wage differential. The latter pay depends on a service obligation, assumed to be three years. The $15,000 accession bonus is included in the wage differential, as is the $2,000 nuclear career accession bonus. Including the $2,000 career accession bonus in the wage differential could lessen the incentive to complete nuclear propulsion training. One approach is to include the $2,000 in the wage differential after completion of YOS 1 or YOS 2, thereby making it contingent on staying. In our example, the wage differential portion of S&I pay is fairly stable at $25,000 to $30,000 between YOS 5 and YOS 25.

Figure 3.3
Notional Example of Nuclear-Qualified Officer S&I Pay Under a Wage Differential Approach

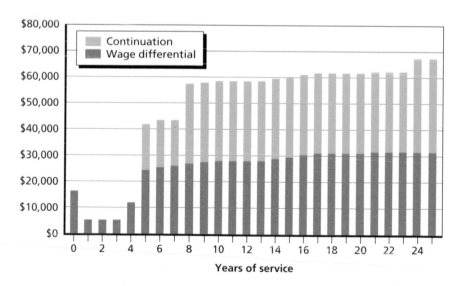

Aviators

Our example is for a Navy pilot. In the Navy, the initial service commitment following initial training as a pilot was eight years, which compares to ten years in the Air Force. Aviation Career Incentive Pay (ACIP) starts at entry into flight training at a monthly rate of $125, which increases to $650 after six years of aviation service, including flight training, and to its maximum of $840 after 14 years, then decreases to $250 after 25 years. Officers who receive ACIP may not receive Hazardous Duty Incentive Pay for flying duty, so it is omitted from the example. A feature of ACIP is that it is for officers who remain in the aviation service on a career basis. As DoD's authoritative guide (Under Secretary of Defense, 2017) states, an officer who completes twelve years in the aviation service including flight training is entitled to continuous ACIP. Officers above O-6 with more than 25 years of aviation service are not entitled to ACIP, however.

In the Navy, the aviator retention bonus is known as Aviator Career Continuation Pay (ACCP; also called Aviator Continuation Pay, ACP). The pilot must be obligated to five years of service to receive ACCP, and the amount of ACCP varies by type of aircraft. The pilot in the example is a fixed-wing fighter attack pilot, a position that pays the highest ACCP at $125,000. The highest Aviator Retention Bonus rate currently permitted by law is $25,000 per year, which amounts to $125,000 for a five-year obligation. The Navy also pays this amount to fixed-wing attack warning pilots and fixed-wing attack electronic pilots, while other pilots are offered ACCP of $75,000 to $100,000 for undertaking an obligation of five years (U.S. Department of Defense, 2016; U.S. Navy Personnel Command, undated). The officer in the example is obligated to receive ACCP through 20 years of service; after 20 years, the pilot chooses an incentive offered by the Navy to commanding officers (O-5) to stay for two years prior to the completion of 22 years of service. This incentive is part of the Navy's ACCP program and is called the Aviator Command Retention Bonus (ACRB), paying $18,000 per year (U.S. Department of the Navy, 2015).

The S&I pays in this example (Figure 3.4) are relatively low in the first eight years of service, as the pilot is under the initial service obligation, but then increase to over $30,000 per year through the

Figure 3.4
Notional Example of Navy Fighter Pilot S&I Pay

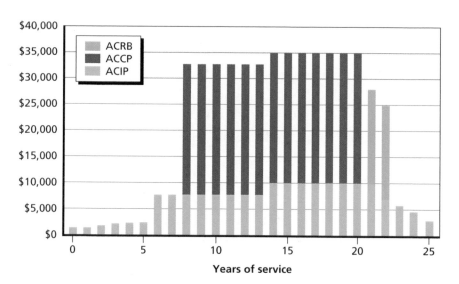

completion of 20 years of service. In YOS 21 and YOS 22, pay declines because of the shift from ACCP to the lower-paying ACRB and the decrease in ACIP.

The S&I pay for the aviator community is a natural example of the wage differential concept because ACIP is clearly a stable and predictable pay that naturally falls into the definition of the wage differential. In contrast, ACCP and ACRB naturally fall outside the definition because they are contingent on a service obligation. Figure 3.5 shows Navy fighter pilot S&I pay under the wage differential approach where ACCP and ACRB are outside the wage differential and ACIP becomes the wage differential.

Conclusion

The examples for internists, nuclear-qualified officers, and aviators illustrate two main points. First, the wage differential concept could easily be implemented for these occupations. As mentioned, the occu-

Figure 3.5
Notional Example of Navy Fighter Pilot S&I Pay Under a Wage Differential Approach

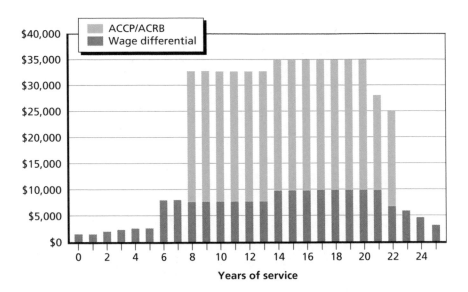

pations chosen for the examples meet the criteria of having high S&I pay, and the S&I expenditures in these occupations accounts for a sizable share of overall S&I pay expenditures. Therefore, the examples reflect the feasibility of implementing a wage differential and targeting it on occupations where it is likely to be a significant component of a servicemember's earnings and make up a noticeable portion of the overall S&I budget. By concentrating on occupations accounting for a relatively high fraction of the S&I budget, the administrative burden of implementing a wage differential would be relatively small compared to the burden if it were implemented to a larger number of occupations. Second, the ease of implementing a wage differential in the example occupations comes from the fact that the S&I pays in these occupations that would be converted to a wage differential already serve the same purpose. As a result, moving to a wage differential would largely be an exercise in relabeling and therefore would be expected to produce little if any gain in the efficiency of compensation or in the value of the compensation package to the servicemember.

CHAPTER FOUR

Dynamic Retention Model Analysis of Alternative Wage Differential Approaches

This chapter considers the cost of various blends of S&I pay to achieve a given retention profile by years of service, where one part of the blend—the wage differential—is paid unconditionally, and the remaining part is structured like a bonus or multiyear special pay where receipt of the pay depends on a service obligation. A blend that puts all the S&I pay in the form of a wage differential that does not depend on a service obligation has less of a retention incentive than a blend that has no wage differential and has all the S&I pay depend on a service obligation. Thus, with all else being equal, we would expect the cost of achieving a given retention profile to be higher when the blend favors the wage differential over obligation-contingent pay. This chapter provides evidence of this, making use of RAND's DRM capability for Air Force aviators.

For three reasons, the Air Force rated community is a useful one to consider as an example of the costs of different blends of S&I pay. First, as was shown in Chapter Three, aviators are among the promising set of occupations for considering the wage differential concept. Second, the Air Force only uses two S&I pays to manage the retention of aviators and to address external market forces that can impact the retention of military aviators, so the Air Force case is relatively simple to consider.[1] These are the incentive pay and the retention bonus. In

---

[1] This advantage of simplicity is offset by a disadvantage in that we are unable to show in the case of U.S. Air Force (USAF) aviators the advantage of rolling up S&I pays that are rela-

the Air Force, these are ACIP (also referred to as Aviator Pay) and Aviator Retention Pay (ARP). Historically, all rated personnel have received ACIP, which has the purposes of compensating for a career that is more hazardous than most military careers and providing a retention incentive. ACIP pays up to $840 a month for midcareer officers. Notably, ACIP is a wage differential that does not depend on a service obligation. ARP is received by Air Force rated personnel who commit to a multiyear obligation, and it typically varies with the occupation and length of the obligation incurred.[2] Finally, RAND recently estimated a DRM of Air Force rated retention that can be used to explore the cost of alternative blends of S&I pay (Mattock et al., 2016).

The DRM is an econometric model that shows how Active Component (AC) and Reserve Component (RC) retention are affected by changes to pay and personnel policies. The model is designed to address questions related to how changing the level and structure of compensation affects retention over a military career, as well as cost in the steady state and in the transition to the steady state. In our work, we estimated a DRM for Air Force rated personnel using pilot cohorts entering from 1990 to 2000 and followed to 2012. We account for changes in military pay since 1990 and changes in ACP in the Air Force that occurred in the first decade of the twenty-first century, the years after these cohorts completed their initial active duty service obligation. In addition, the model incorporates the pilot's ACP/contract length choice and includes major airline hiring and the unemployment rate, as well as the availability of ACIP.

We use the DRM for Air Force rated personnel to conduct simulations of achieving a steady state baseline retention profile by years of service using different blends of ACIP (the wage differential) and ACP. The baseline steady state retention profile is the one predicted under current ACIP and ACP policy. We then consider the cost of achiev-

---

tively constant and do not depend on a service obligation, since the USAF rated community only has one such pay, ACIP, and not multiple such pays.

[2]   Three common options that have been offered by the Air Force have been a three-year contract, a five-year contract, and a contract until 20 years of aviation service at amounts that are now up to $35,000 per year and previously were up to $25,000 per year. ACP maxima have changed over time, and a history of ACP is provided in Mattock et al. (2016).

ing this retention profile through two scenarios, under two alternative blends of S&I pay: (1) ACIP and no ACP; and (2) ACP and no ACIP. The DRM simulation capability includes an optimization routine that finds the level of ACIP in the first scenario or the level of ACP in the second scenario that would sustain retention relative to the baseline and computes the change in cost. Scenario 1 corresponds to the case in which S&I pay is only in the form of a wage differential (in this case, ACIP), while scenario 2 represents a case in which S&I pay has no wage differential and is only in the form of pay contingent on a service obligation. Once the optimized value is found, the simulation then computes the cost of USAF ACIP and ACP in total under the baseline and under each scenario.

The purpose of considering these extreme scenarios—no ACP versus no ACIP—is not because they are realistic scenarios or ones that we would recommend; rather, our purpose is to illustrate that a different blend of S&I pay has retention effects, and these retention effects have cost implications. The extreme cases show the upper and lower bounds for setting the wage differential, the additional cost of rolling into the wage differential a part of S&I pay that currently depends on a contract obligation, and the savings of reducing the wage differential and rolling part of the wage differential into a part that depends on an obligation.

We next present results of the DRM analysis. As supporting material, Appendix A presents an overview of the DRM and a discussion of how we compute cost. A more detailed presentation of the USAF DRM is given in Mattock et al. (2016).

Results

Figure 4.1 shows the simulated steady state retention profile for USAF rated personnel in the baseline versus each scenario. Scenario 1 (ACIP only) is at the top, and scenario 2 (ARP only) is below it. The baseline (the black line) is the predicted cumulative probability of retention at each year of service under current USAF ACP and ACIP policy. The red line is the predicted cumulative probability under each scenario. Table 4.1 shows information relevant to each scenario.

Figure 4.1
USAF Rated Officer Retention Under Scenarios 1 and 2

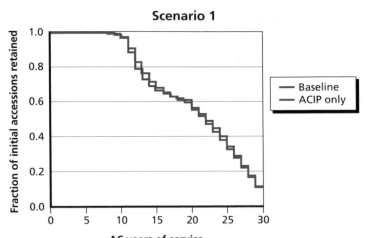

Scenario 1

Change in force, 0%; change before 20 years, 0.4%;
change after 20 years, −3.8%

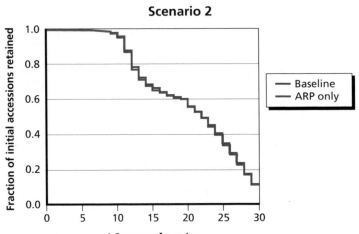

Scenario 2

Change in force, 0%; change before 20 years, −0.4%;
change after 20 years, 1.9%

Table 4.1
Steady-State ACIP and ACP Costs, 2014 Dollars

	Percentage Change in ACIP to Sustain Retention	Percentage Change in ACP to Sustain Retention	Baseline Cost	Policy Scenario Cost	Difference	Percentage Difference
Scenario 1 (All ACIP)	264%	NA	$64,203,370	$170,879,300	$106,675,930	166.2%
Scenario 2 (All ACP)	NA	27%	$64,203,370	$33,184,650	−$31,018,720	−48.3%

Under scenario 1, ACP is eliminated. The retention profile at the top in Figure 4.1 shows that eliminating ARP and replacing it with ACIP can closely but not quite replicate the USAF rated retention profile. Midcareer retention between YOS 10 and YOS 15 is somewhat higher than the baseline, while retention after YOS 20 is slightly lower. It would be possible to improve the fit relative to the baseline if we changed the years of service that are targeted under ACIP and could vary the dollar amounts. In contrast, as shown below it, it is possible to produce almost the identical retention profile as the baseline when ACIP is eliminated and replaced with ARP.

We find that ACIP, the wage differential portion of S&I pay, would have to increase by 264 percent to sustain retention relative to the baseline in the steady state. On the other hand, we find that ARP would have to increase by 27 percent to sustain retention.

Table 4.1 shows that baseline ARP and ACIP costs for the USAF rated community are $64.2 million in 2014 dollars. Under scenario 1, these costs increase to $170.9 million, or by 166 percent, more than doubling the cost. The reason for the cost increase is that eliminating ARP means that S&I pay no longer depends in part on a service obligation, so the retention effect of the S&I pay decreases. ACIP would have to increase by more than a dollar-for-dollar rate when ARP is eliminated because ACIP does not entail a service obligation. The table also shows the other extreme case, eliminating ACIP. In this case, costs decrease to $33.2 million, or by 48 percent, implying that a blend of

S&I pay that favors making S&I pay contingent on a contract obligation is more cost-effective.

More generally, the results imply that a blend of S&I pay that favors a wage differential and rolls into the wage differential all or part of S&I pay that depends on a contract obligation will increase costs.

CHAPTER FIVE

Disbursing Special and Incentive Pay When Uncertainty Is Present

Special and Incentive Pay as Insurance

When uncertainty exists, the military can either compensate in advance for the expected loss when adverse circumstances are realized or, alternatively, insure the servicemember against adverse realizations by committing to a rate of reimbursement conditional on the realization. If the servicemember is risk averse, we show that it costs the military less to offer this insurance than to pay an ex ante amount that would provide the servicemember with the same expected utility.[1] This provides a rationale for using S&I pay to compensate for adverse circumstances when they occur rather than providing higher pay up front—as in the form of a wage differential that would be paid regardless of military circumstances such as assignment or hazardous duty and regardless of supply-and-demand conditions.

To see this, suppose the servicemember's utility depends on wealth w and the state of the world, which is low l with probability q and high h with probability $1 - q$. Expected utility is

---

[1] This point is distinct from the notion in the theory of compensating differentials under which (1) individuals will sort themselves according to occupation riskiness and personnel risk preference, with the most risk tolerant individuals going to the riskiest occupations; and (2) in equilibrium, compensating differentials may exist across occupations, depending on the supply of individuals with risk tolerance and the demand for people with that trait. If they do exist, the differential is higher in riskier occupations. The idea of compensating differentials for military combat occupations is discussed in Simon et al. (2012).

$$q\,U(w,l)+(1-q)U(w,h)$$

Additionally, consider insurance that paid an amount δ upon the realization of the low state, such that utility in the low state would be the same as utility in the high state: $U(w+\delta,l)=U(w,h)$. That is, δ is the minimum amount the servicemember would be willing to accept to be held harmless against the low state compared to the high state.

The expected cost of providing this amount is the probability of the low state times the amount of the payment: $q\delta$. Now, suppose the insurance were not provided and ask, by how much could wealth in the high state be decreased yet provide the same utility as expected under the uncertainty of the high and low outcomes? This would be an amount κ such that

$$U(w-\kappa,h)=qU(w,l)+(1-q)U(w,h)$$

In effect, κ is the maximum amount the servicemember would be willing to spend on insurance against the risky outcome. The servicemember is not explicitly buying insurance though; it is being provided by the military. The servicemember's wealth is related to the total amount of military pay received, and we can interpret κ as the amount by which the military could decrease the servicemember's wealth (e.g., by decreasing military pay) and still leave the servicemember as well off as under the uncertain outcome. If the military can save more ex ante (κ) than it expects to pay out under the low outcome ($q\delta$), then it is cost-effective to have a policy of making payment conditional on the low outcome. This will hold if $\kappa \geq q\delta$.

To show that this is the case, suppose $q\delta$ were taken away instead of κ. With risk aversion, the utility function is concave, and for a concave function Jensen's inequality implies that $U(w-q\delta,h)\geq qU(w,l)+(1-q)U(w,h)$. But since utility is increasing in wealth and $U(w-\kappa,h)=qU(w,l)+(1-q)U(w,h)$, it follows that $\kappa \geq q\delta$. Therefore, a policy of, in effect, decreasing military pay by κ relative to what it would have been if the risk of a low outcome were 0, and paying δ upon realization of the low outcome with probability q, is cost-effective.[2]

The disutility of the low state is not fixed but random, and the military establishes the S&I pay rate for the low state ex ante. This means that a pay rate of δ might not compensate for the realized conditions in the low state, or $U(w+\delta,l) < U(w,h)$. In that case, additional S&I pay may be needed to sustain retention, and as mentioned bonuses can be turned on or increased to provide the additional pay. If the realized conditions in the low state are better than expected, $U(w+\delta,l) > U(w,h)$, and retention will be sustained or increase.

Setting the optimal level of S&I pay to be disbursed when the low state is realized in principle involves the risk distribution, the risk tolerance distribution of servicemembers, the cost of S&I pay δ, plus any bonus needed to sustain retention. A challenge is that δ is usually set ex ante by DoD or by Congress before the specific nature of the risk is known—that is, the ex ante commitment to pay δ is made in the form of a special pay, but the precise circumstances that will be realized in the low state are not known ahead of time. The rate δ must be high enough such that, given beliefs about the risks faced in the low state, $q\delta$ will be sufficient to compensate the member at the margin of staying in service or leaving. Being able to pay bonuses on top of δ if servicemember low-state realizations are worse than expected provides the military with flexibility to allocate additional pay needed to maintain retention. Also, Congress and DoD may act to change δ. Hostile Fire/ Imminent Danger Pay was increased in 2003 to $225 per month for servicemembers deployed in Afghanistan and Iraq, for example (Gould and Horowitz, 2012).

S&I pay for assignment, special duty, and hazardous duty are defined as monthly rates. This implies that the servicemember is insured in proportion to the duration of the assignment or duty. The assignment and special duty pay rates depend on the specific assign-

---

[2]   Subramaniam (2016) uses a similar approach to argue that medical innovations should not be valued solely on patients treated with the innovation but should also include the insurance value to all prospective patients from knowing that a new treatment is available. Ignoring the latter would lead to under investment in medical innovation.

ment and type of special duty, but there is a single pay rate for hostile fire. In effect, that assumes that the pay rate for the servicemember at the margin is the same for all hostile deployment/imminent danger situations. However, junior enlisted personnel proved to be at higher risk of becoming wounded or killed in action from 2005 to 2010 (U.S. Department of Defense, 2008). Therefore, it is not surprising that during this period the military made greater use of bonuses to stabilize retention. There were also differences in deployment rates by occupation, which might suggest that δ should differ by occupation. But if servicemembers sort themselves into occupations according to their risk preferences, different δs by occupation may not be necessary. This would be the case if occupations with high risk attracted individuals with low risk aversion. Thus, servicemembers in different occupations and subject to different risks might be satisfied with a common rate δ.[3]

Value to the Servicemember if Special and Incentive Pay Uncertainty Were Eliminated

The previous section reasons that S&I compensating for uncertain circumstances should not be included in a wage differential because it is not cost-effective to do so. This section sets aside the issue of cost and asks what eliminating variation in pay resulting from uncertainty would be worth to the servicemember. In particular, we use a formula for the *certainty equivalent*, which is the amount a risk-averse individual is willing to pay to avoid risk. We draw on earlier research on military cash compensation that provides an estimate of bonus variation relative

---

[3] Finkelstein and McGarry (2006) make a different but related point with respect to self-selection into insurance markets. Although Rothschild and Stiglitz (1976) found theoretically that even a small amount of adverse selection could destabilize insurance markets, the markets are typically stable in practice. Finkelstein and McGarry argue that selection into insurance markets is both adverse and advantageous, with advantageous selection a stabilizing force. Adverse selection occurs when individuals given private information that they are riskier than observationally equivalent individuals are more likely to buy insurance. Advantageous selection occurs when individuals who are less risky than other observationally equivalent individuals are more likely to buy insurance. Relating this to the military, the individual's preference is the private information.

to total cash compensation by year of service and use it to compute the certainty equivalent.

We present an estimate of the certainty equivalent for an individual just entering the military as an enlistee and planning to stay for ten years. We assume the individual signed up for an initial term of three years and might have received an enlistment bonus. Whatever its amount, it is fixed during the term and has no uncertainty. However, from the perspective of the individual looking forward over a ten-year horizon, there is uncertainty in the bonuses that may be available at first- and second-term reenlistment. The ten-year period is chosen because most reenlistment bonuses are paid during this period (Asch, Hosek, and Martin, 2002). For the certainty equivalent that is derived in Appendix B, the calculation uses the formula

$$ce = \frac{1}{2}\frac{\sigma^2}{w}$$

where σ^2 is reenlistment bonus variance and w is military cash compensation, both per year. Our estimate should be recognized as an upper limit on the certainty equivalent because, as the dynamic retention framework emphasizes, individuals can reoptimize in every period and, even though initially planning to stay for ten years, might choose to leave before that time.

Table 5.1 shows the certainty equivalent as a percentage of the present discounted value of military pay over the first ten years of service. The calculation assumes that the entrant has a three-year obligation, and the enlistment bonus, if any, is known at the time of entry.

Table 5.1
Certainty Equivalent over the First Ten Years of Service as a Percentage of the Present Discounted Value of Military Pay

Air Force	Army	Marine Corps	Navy
0.15	0.02	0.00	0.34

SOURCE: Authors' calculations based on 1999 military pay data. See Appendix B for details.

Thus, there is no bonus uncertainty in the first three years of service. Appendix B has the details of the calculation.

The certainty equivalent estimate for the Air Force is 0.15 percent of the present discounted value of pay over the first ten years of service. The estimates for the Army and Marine Corps are lower, reflecting little use of reenlistment bonuses by these services in 1999, and the Navy estimate is just over one-third of 1 percent. The results are based on historical data and should not be interpreted as reflecting current bonus usage by the services, but instead are illustrative and show a range of estimates based on differences in bonus usage.

The certainty equivalents shown in Table 5.1 are quite small because the standard deviations of reenlistment bonuses relative to military pay are quite small. The results indicate that even if bonus variation were twice as high as the Navy estimate, the largest in the table, the certainty equivalent would be about two-thirds of 1 percent of the present discounted value of pay over the first ten years of service. Similarly, if the certainty equivalent formula were doubled to be at its maximum value consistent with published studies, rather than at the average value used for the table, the Navy certainty equivalent would then be about 1.33 percent of the present discounted value. Again, these values are upper limits because the entrant is assumed to stay in the service for ten years, whereas the individual could reoptimize in future periods and could choose to leave before completing ten years. The results suggest that the value to the servicemember of eliminating reenlistment bonus uncertainty would be small.

Finally, the certainty equivalent is distinct from the value to the service of being able to use bonuses to sustain retention and regulate personnel flows. Replacing bonuses with a wage differential would decrease the armed services' flexibility to respond when conditions hurt retention or when retention needs to be increased, and thus could hurt military capability. Today the services can vary the availability and amount of accession and retention bonuses, and, for example, they used this flexibility to increase bonuses during extensive operations in Afghanistan and Iraq (Hosek and Martorell, 2009) and to decrease them during the Great Recession when civilian employment opportunities fell. For officers, bonuses in health professions, aviation, and for nuclear qualification are capped, and the caps cannot be changed at the discretion of the armed services; Congress can, however, change the caps. For example, when the Air Force faced difficulty in retaining pilots as major airlines' demand increased, Congress acted to increase the cap on pilot retention bonuses.

Conclusion

This chapter addressed two aspects of military pay—namely, whether to compensate for unpredictable circumstances such as deployment on an ex ante basis or, alternatively, when they are realized, and whether servicemembers could expect a substantial increase in the value of their military compensation package if uncertainty related to S&I pay were eliminated, as would be the case if the S&I pay were replaced by a wage differential. We used economic theory to derive the finding that it is more efficient to compensate for unpredictable circumstances when they are realized rather than to compensate ex ante. The finding assumed that servicemembers were averse to risk and further assumed that the compensation policy—paying when unpredictable circumstances were realized—was common knowledge and a credible commitment (i.e., servicemembers knew about it, and the military would make the payment when the conditions for it arose). With respect to eliminating the uncertainty of S&I pay by converting to a wage differential, we used expected utility theory and estimates from prior research to estimate the value to servicemembers of this change. We found the value to be small relative to total compensation. We noted, too, that changing to a wage differential would decrease the armed services' flexibility to respond to unpredictable circumstances.

CHAPTER SIX

Conclusion

The Office of Compensation within the Office of the Under Secretary of Defense for Personnel and Readiness asked RAND to conduct research into the concept of a wage differential. Under this concept, military occupations could receive additional compensation on a steady basis, such as a schedule based on occupation or duty, pay grade, and years of service. A potential advantage is that a wage differential would provide greater stability in military compensation.

The steps in our research included categorizing S&I pays by four attributes: (1) whether the pay provides an occupational differential; (2) whether the pay contains an incentive to select a longer service obligation; (3) whether the pay compensates for hazardous duty or language proficiency; and (4) whether the pay compensates for military circumstances such as hostile deployment, assignment, mission, location, or tempo.

Our key findings were, first, that S&I pays that include an incentive to select a longer obligation are far more cost-effective than pays without that feature. In other words, a blend of S&I pay that favors making S&I pay contingent on a multiyear service obligation can be a more cost-effective means of achieving retention goals than a blend that favors a fixed wage differential. Second, on theoretical grounds there is reason to use S&I pays as insurance against uncertain outcomes, making payments when adverse conditions are realized rather than including them in a wage differential that would be paid independently of such outcomes. It is more cost-effective to pay reenlistment bonuses when external employment opportunities can decrease retention or to disburse Hostile Fire/Imminent Danger Pay when personnel are deployed, for example, than to pay a scheduled amount regardless of those conditions. Third, although pay that is conditional on circumstances is uncertain because of uncertainty about when such circumstances will arise, the value to the servicemember of eliminating the uncertainty in pay appears to be small. In contrast, the value to the armed services of using S&I pays to respond to circumstances when they arise may be large because they help to sustain retention and force capability. Fourth, a number of S&I pays are so-called incentive pays; these are stable, scheduled pays. In effect, they are wage differentials.

Finally, to gain some sense of the extent to which wage differentials are already present in the form of incentive pays, we analyzed S&I pays among Navy officers. We found that S&I pay predominates in three occupational fields: health care services, aviation, and naval operations, the last of which includes nuclear-qualified officers. Appendix C describes the analysis and presents the results for all 1,836 four-digit occupations, though much of the S&I expenditure is concentrated in the top 400 occupations ranked by average total S&I pay. Table 6.1 presents the results for these occupations. As can be seen, the fields of health care services, aviation, and naval operations account for over one-third of Navy officers—36 percent—but over two-thirds of overall S&I expenditures on Navy officers—70 percent.

We know from Chapters Two and Three that health professions officers, aviation-related officers, and nuclear-qualified officers have sizable accession bonuses, incentive pay, and retention bonuses. A wage differential initiative focused on these occupations would thus cover a significant fraction of officers receiving S&I pays, and they would therefore be among the most affected by a wage differential. Yet the incentive pay in these occupations serves as a wage differential. At the same time, the retention bonuses in these occupations lends greater cost-effectiveness than would be achieved if all S&I pay in these occupations were made into a wage differential.

Overall, our research provides new insights into the structure and cost-effectiveness of existing S&I pays. We find that a wage differential is already present in many S&I pays, while other pays, which are paid conditional on circumstances, are more cost-effective in their cur-

Table 6.1
Distribution of S&I Pay Expenditure and Distribution of Navy Officers for One-Digit Occupations and for 400 Top Four-Digit Occupations Ranked by Average S&I Pay

Field Number	Occupation Field	NOBC Range	Number of Four-Digit Occupations	Number of Officers	Share of S&I Expenditures	Share of Officers
0	Health care services	1–999	85	15,813	36.3%	10.8%
1	Supply and fiscal	1000–1999	13	388	0.1%	0.1%
2	Sciences and services	2000–2999	30	1,952	1.2%	0.8%
3	Personnel	3000–3999	31	6,380	3.2%	3.0%
4	Facilities engineering	4000–4999	1	1	0.0%	0.0%
5	Electronics engineering	5000–5999	7	81	0.0%	0.0%
6	Weapons engineering	6000–6999	10	199	0.1%	0.1%
7	Naval engineering	7000–7999	17	1,812	0.9%	0.9%
8	Aviation	8000–8999	77	14,291	10.9%	7.5%
9	Naval operations	9000–9999	129	37,857	22.8%	18.1%
	Totals		400	78,774	75.7%	41.3%

rent form than if they were if paid as a wage differential. Finally, the findings indicate greater cost-effectiveness when S&I pay includes an incentive to select a longer obligation, which therefore suggests going in the direction of making greater use of such incentives rather than in the direction of a wage differential.

Overview of the Dynamic Retention Model

The foundation of the DRM is a theory of retention decisionmaking over a servicemember's career. The theory is a mathematical model of individual decisionmaking in a world with uncertainty and where individuals are heterogeneous in terms of their tastes for military service. The model begins with service in the AC, and individuals make a decision to stay or leave each year. Those who leave the AC take a civilian job and at the same time choose whether to participate in the RC. The decision of whether to participate in the RC is made each year, and individuals can move into or out of the RC from year to year. More specifically, a reservist can choose to remain in the RC or to leave it to be a civilian, and a civilian can choose to enter the RC or remain a civilian.

The DRM for USAF rated personnel also includes the pilot's choice of a multiyear contract under the ARP program. Rated personnel who choose a longer contract receive ARP for more years, but they are also locked in to their contract so that they forgo the opportunity to take advantage of better opportunities that might present themselves during the contract period. This multiyear contract length choice is modeled as a nested choice made under uncertainty. The uncertainty arises from not knowing the specific future conditions (e.g., assignments, flying time, deployments) that accompany these choices. The incorporation of this nested ARP contract length choice requires estimation of an additional parameter in the model, related to the variance of the shock associated with the multiyear contract choice. (This parameter estimate is statistically significant, indicating that this portrayal of the multiyear contract choice is an improved approach to modeling the ARP/contract choice.)

The parameters of this model are empirically estimated with data on Air Force rated careers drawn from administrative data files, specifically the Defense Manpower Data Center (DMDC) Work Experience File, which contains person-specific longitudinal records of AC and RC service. We use the Work Experience File data for servicemembers who began as Air Force officers between 1990 and 2000, considering each entry cohort separately and tracking their individual careers in the AC and the RC, if they join the RC, through 2012. By considering multiple entry cohorts, we are able to incorporate changes in USAF ARP policy that occurred in the first decade of the twenty-first century and changes in military pay since 1990. For each officer in an entry cohort, we constructed his or her history of AC and RC participation and used these records in estimating the model. We supplement these data with information on active, reserve, and civilian pay. Specifically, this incorporates features of the civilian pilot opportunities available to pilots who leave the Air Force; it is garnered from a regression estimation of veteran pilot earnings and uses an expected wage line that is a combination of veteran civilian nonpilot and veteran civilian pilot earnings, where earnings are weighted according to an estimated probability that officers are hired by a major airline, where the probability of being hired is a function of the number of major airline hires in a given year.[1] We estimate the DRM using maximum likelihood methods and find all parameter estimates to be statistically significant.

A limitation of this model is that the Air National Guard and Air Force Reserve are not treated separately but are combined into a single group, USAF RC. Additionally, the model excludes demographic variables such as gender, marriage, and spousal employment. We also

---

[1] In the simulations shown in Chapter Five, the probability of being hired by a civilian airline is set to 0.10 under both the baseline and each policy scenario. And because the focus of this analysis is not on the effects of civilian opportunities on pilot retention (as in Mattock et al., 2016), we assume a 0 percent change in civilian pilot pay and a 0 percent change in civilian nonpilot pay.

exclude health status and health care benefits, and we do not explicitly model deployment or deployment related pays.

That said, the estimated models fit the observed data extremely well for the both the AC and the RC. Once we have parameter estimates, we can then use the logic of the model and the estimated parameters to simulate the AC cumulative probability of retention in each year of service in the steady state for a given policy environment, such as an increase in ARP or a change in the level of ACIP or an increase in the civilian opportunity wage facing Air Force pilots. The simulation output includes a graph of the AC retention profile by years of service. As has been discussed in Mattock et al. (2016), we consider how well the model fits the observed data by simulating the steady-state retention profile in the baseline—or current policy environment—and comparing the simulation to the retention profile observed in the data. As mentioned, we find that the model fit is good, and the model prediction does not deviate far from the observed data.

As part of the simulation capability, the DRM computes optimized values of ACIP or of ARP, depending on the scenario, to sustain USAF rated retention relative to the baseline. This involves computing the value of either ACIP or ARP that minimizes the difference between the baseline retention profile under the USAF current ACIP and ARP policy and the profile under the scenario being considered. The optimized value of ACIP in scenario 1 and of ARP in scenario 2 is the value that does best in keeping the retention profile constant relative to the baseline.

In addition to computing the optimized values, the simulation also computes ARP and ACIP costs in the baseline and under each policy scenario. Of course, there are other elements of personnel costs, but because the retention profile is being held constant, these are unchanged and so are not included in the computation of the change in costs under the policy scenario.

In short, the DRM is firmly grounded in the theory of retention decisionmaking and empirically grounded in data on the actual retention behavior of USAF rated officers over a more than 20-year period for the earliest entry cohorts. Further, it includes a simulation capability that allows assessment of compensation changes that have yet to be tried or of policy changes for which it would be difficult to get a counterfactual of what would have happened in the absence of the policy change. That is, it permits "what if" analyses of changes in compensation policy, even for changes that may be outside of historical experience, such as alternative blends of S&I pay.

Certainty Equivalent Calculation

Risk Aversion

When there is diminishing marginal utility and income is random around a mean, the expected utility of income is less than the utility of the mean income. The difference between the expected utility of income and the utility of the mean income is a measure of the extent of risk aversion.[1] If the utility function has more curvature around the mean, this difference will be greater. This suggests that risk aversion depends on the curvature of the utility function, and curvature depends on the rate of change in marginal utility relative to marginal utility. These insights underlie the Pratt-Arrow measure of absolute risk aversion:

$$R(w) = -u_{ww}/u_w$$

The measure in general depends on preferences for wealth and can change as wealth changes. For instance, at high levels of wealth its marginal utility might be low, though at the same time there might be little change in marginal utility with respect to wealth. Recognizing the dependence on wealth, another measure of risk aversion is relative risk aversion:

$$R_r(w) = wR(w)$$

---

[1] For a general discussion of risk aversion, see Varian (1992).

Certain utility functions have constant relative risk aversion (CRRA). This can be a useful simplifying assumption when the changes in wealth being considered are relatively small and it is reasonable to choose a specification where the measure of risk aversion is independent of wealth. For instance, a CRRA utility function has the form

$$u(w) = \frac{w^{1-\alpha}}{1-\alpha} \text{ for } \alpha > 0 \text{ but } \alpha \neq 1$$

$$u(w) = \ln(w) \text{ for } \alpha = 1$$

In this case, the coefficient of relative risk aversion is α.

Certainty Equivalent

The certainty equivalent (*ce*) is the amount that an individual is willing to pay to avoid risk. It can be thought of as the amount by which military pay can be reduced when risk is eliminated, such that the utility of certain pay equals the expected utility of risky pay. In a one-period model, *ce* is defined by the relationship

$$u(w - ce) = E[u(w + \varepsilon)]$$

where ε is a random term with an expected value of 0 and E is the expectation operator. Implicitly, the random term is measured in the same units as pay. For small ε, a Taylor approximation of the left-hand side around ε and *ce* = 0 gives

$$u(w - ce) \cong u(w) - ce\ u_w$$

A Taylor approximation of the right-hand side around ε is

$$E[u(w + \varepsilon)] \cong E[u(w) + \varepsilon u_w + \frac{1}{2}\varepsilon^2 u_{ww}] = u(w) + 0 + \frac{1}{2}\sigma^2 u_{ww}$$

where σ^2 is the variance of the random component. Equating the approximations gives

$$ce = -\frac{1}{2}\sigma^2 u_{ww}/u_w = \frac{1}{2}\sigma^2 R(w)$$

With positive and diminishing marginal utility of w, $u_w > 0$, $u_{ww} < 0$, ce is positive. Also, ce is positively related to the variance of the random term. Intuitively, the larger the variance, the smaller will be expected utility and the larger will be the difference between the utility of expected pay and the expected utility of risky pay. Hence, the individual would be willing to give up more pay to avoid the greater riskiness.

From the result for the CRRA utility function, we have

$$ce = \frac{1}{2}\sigma^2 R(w) = \frac{1}{2}\sigma^2 \frac{\alpha}{w}$$

Estimate of α

Chetty (2006) recognizes that labor supply response to wage changes reveals information about the rate at which the marginal utility of consumption diminishes. He shows that the uncompensated wage elasticity of labor can be used to bound risk aversion in an expected utility model. He examined 13 peer-reviewed empirical studies of labor supply, which in general find that the uncompensated wage elasticity of labor supply is not very negative (an exogenous wage increase brings about a small decrease in labor supply). Using these estimates, Chetty inferred a mean estimate of the coefficient of relative risk aversion of 1 for a CRRA utility function, implying log utility, and an upper bound of 2. Values above 2 require that wage increases cause a greater decrease in labor supply than is found in studies.

In the case where $\alpha = 1$, which is Chetty's mean estimate of α, the ce of CRRA utility is

$$ce = \frac{1}{2}\frac{\sigma^2}{w}$$

Calculation

Table B.1 shows military pay (cash compensation) and the bonus-related standard deviation in pay, approximated from figures in Asch, Hosek, and Martin (2002).[2] Military pay is higher today than when that study was done, but its increase with respect to years of service is similar. The table shows the bonus standard deviation by service, which we use to show the sensitivity of the certainty equivalent estimates to a range of bonus standard deviation.[3]

Table B.1
Military Pay and Reenlistment Bonus Standard Deviation, 1999 (in 1999 dollars)

YOS	Pay	Standard Deviation			
		Air Force	Army	Marine Corps	Navy
0	22,000	2,000	800	0	3,000
1	23,300	2,000	800	0	3,000
2	24,600	2,000	800	0	3,000
3	25,900	2,000	800	0	3,000
4	27,200	2,000	800	0	3,000
5	28,500	2,000	800	0	3,000
6	29,800	2,000	800	0	3,000
7	31,100	2,000	800	0	3,000
8	32,400	2,000	800	0	3,000
9	33,700	2,000	800	0	3,000
10	35,000	2,000	800	0	3,000

SOURCE: Asch, Hosek, and Martin, 2002.

---

[2] We use the figures because the underlying data are not available.

[3] Note that the bonus standard deviation will scale up by the same percentage as basic pay has grown over time. This is because reenlistment bonuses have been based on basic pay.

We apply the formula given above with $\alpha = 1$ to calculate the certainty equivalent by years of service (see Table B.2). In the first three years of service, the certainty equivalent is 0. This is because the entrant is assumed to have an initial obligation of three years and knows the amount of the enlistment bonus, if any; it is not uncertain. From the perspective of the entrant, there is bonus uncertainty at first- and second-term reenlistment, and this gives rise to the certainty equivalent shown for YOS 4 through YOS 10. The certainty equivalent decreases as years increase. This is because military pay increases while the bonus standard deviation remains constant.

The final step is to compute the present discounted value of pay and the certainty equivalent (see Table B.3). We use a discount factor of 0.90 based on the discount factors for enlisted members estimated by Asch, Mattock, and Hosek (2017). As can be seen, the certainty equivalent is less than 1 percent of the present discounted value of pay. For instance, the Air Force percentage is 0.15 percent.

Table B.2
Certainty Equivalent of Reenlistment Bonus, 1999 (in 1999 dollars)

| YOS | Standard Deviation | | | |
	Air Force	Army	Marine Corps	Navy
0	0	0	0	0
1	0	0	0	0
2	0	0	0	0
3	77	12	0	174
4	74	12	0	165
5	70	11	0	158
6	67	11	0	151
7	64	10	0	145
8	62	10	0	139
9	59	9	0	134
10	57	9	0	129

Table B.3
Present Discounted Value of Military Pay to YOS 10 and Certainty
Equivalent of Reenlistment Bonus, 1999 (in 1999 dollars)

YOS	Military Pay	Certainty Equivalent			
		Air Force	Army	Marine Corps	Navy
Present discounted value	186,371	282	45	0	634
Percentage		0.15	0.02	0.00	0.34

NOTE: Discount factor = 0.90.

Special and Incentive Pay Concentration: Navy Officers

This appendix illustrates an approach to determining the concentration of S&I pay. A reason for considering concentration is to learn whether a wage differential initiative that focused on a limited set of occupations would nevertheless encompass significant fractions of personnel and of total S&I expenditures. This would be the case if S&I pay were concentrated. Equally important is to learn whether the more concentrated occupations had S&I pay that included an incentive to select a longer obligation, as it would not be cost-effective to roll it into a wage differential (see Chapter Four).

We apply the approach to Navy officers, though it could be applied to other groups. Using data described below, we first find occupations with high average total S&I pay and then, sorting occupations from highest to lowest, we determine whether these occupations also account for a large share of S&I expenditure. We briefly describe our database, method of inferring total S&I pay, and occupational classification.

Database

Our data file consists of individual records for 2002–2012 linked from the following sources:

- the DMDC Active Duty Pay File, which includes data on basic pay, allowances, and some but not all types of S&I pay

- the Defense Finance and Accounting Service (DFAS) for data on total cash compensation
- the Defense Enrollment Eligibility System Point-in-Time Extract file for individual demographic data
- the DMDC Work Experience File for occupational histories and job characteristics.

Linking was done with a scrambled Social Security number provided by DMDC, and all data were privacy protected.

Data on S&I Pay

The DMDC Active Duty Pay File provided 21 S&I pay fields. Table C.1 lists the types of S&I pay by the categories used in the pay file. The categories are for nuclear-qualified professions, aviation, health professions, hazardous duty, assignment and special duty, and skill/proficiency. The 21 fields include three for hazardous duty pay and two for Foreign Language Proficiency Pay. The data omit some types of S&I pay, so the item total of the pay fields is likely to be less than a servicemember's total S&I pay. However, we were able to infer total S&I pay from DFAS data, as explained below.

We estimated a servicemember's total S&I pay by subtracting basic pay and allowances (DMDC data) from total cash compensation (DFAS data):

$$S\&I \ pay = total \ pay - (basic \ pay + allowances)$$

This approach addressed another limitation of the DMDC data, which is that they contained taxable compensation, not total compensation. Taxable compensation differs from total compensation because of pretax retirement contributions and combat zone tax exclusions. Using taxable compensation would have undercounted compensation.

Table C.1
Special and Incentive Pay as Categorized in the DMDC Active Duty Pay File

Category	Pay
Nuclear	Nuclear Officer Accession Bonus
	Nuclear Officer Career Accession Bonus
	Nuclear-Qualified Officer Continuation Pay
	Nuclear Career Annual Incentive Bonus
Aviation	Aviation Officer Continuation Pay
	Aviation Officer Career Incentive Pay
Health professions	Officer Board Certification Pay
	Medical Officer Retention Bonus, Multiyear Special Pay, or Nurse Bonus
Hazardous duty	Hostile Fire/Imminent Danger Pay
	Hazardous Duty Incentive Pay I
	Hazardous Duty Incentive Pay II
	Hazardous Duty Incentive Pay III
Assignment and special duty pay	Career Sea Pay
	Career Sea Pay Premium
	Enlisted Hardship Duty Pay
Skill/proficiency pay	Diving Duty Pay
	Foreign Language Proficiency Pay I
	Foreign Language Proficiency Pay II
	Enlisted Proficiency Pay
Enlisted	Selective Reenlistment Bonus
	Regular Reenlistment Bonus

SOURCE: DMDC Active Duty Pay File.

Occupation Codes

We use Navy Officer Billet Classifications (NOBCs) in our tabulations; these are four-digit codes that identify a group of similar billets, though not necessarily billets identical in scope and nature of duties. Our data contained 1,835 NOBCs. According to NOBC documentation, NOBCs serve the dual purpose of identifying officer billet requirements and officer occupational experience acquired by serving in a billet or through a combination of education and billet experience (U.S. Bureau of Naval Personnel, 2010). Simply put, NOBC codes reflect duty occupation rather than primary or secondary occupation.[1] The first digit of an NOBC identifies the occupational field, and the second digit identifies the group within the field. The third and fourth digits give the specific billet classification within the group. An officer's NOBC changes as billets change.

To prepare data for analysis, we removed cases with missing NOBCs, missing federal wage variables, duplicate Social Security numbers, or negative pay. This resulted in a reduction of about 16 percent. We adjusted dollar amounts to 2012 dollars using the Consumer Price Index for Urban Consumers, which represents 90 percent of the U.S. population.

Distribution of S&I Pay Across Occupations

Figure C.1 shows the distribution of average total S&I pay by NOBC occupation. The x-axis shows the rank of each four-digit occupation in terms of average total S&I pay, highest first, where the average is the total annual S&I pay across all officers and across all years in our data divided by the number of instances where an officer received S&I pay in that occupation.[2] S&I pay for each officer in each occupation is defined according to the equation immediately above, so it is total S&I pay for that officer. Thus, the x-axis is the occupation's rank. There are

---

[1] For Navy officers, primary occupation codes consist of a four-digit officer designator and the most current three-digit additional qualification designators. Secondary occupation codes are four digits and come from the most current subspecialty code, with a last digit specifying the level of education, training, or experience.

[2] Thus, the average was taken over officer years in the occupation. Because the data are longitudinal, this meant that an individual officer's S&I pay would have been included for as many years, 2002–2012, as the officer was in the occupation and received S&I pay. This does not pose a problem because our interest is in the average S&I pay per year in the occupation, and all officers in the occupation in each year should be counted in the average.

1,835 occupations, and the highest is ranked number 1 and the lowest is 1,835. The *y*-axis on the left side of the chart is average S&I pay. The blue line plots this average by its occupation rank. The top-ranked occupation, anesthesiologist, has average S&I pay of $118,678. Health professions comprise virtually all the occupations ranked in the top 50.[3] Average S&I pay is about $46,000 for the 50th-ranked occupation, $28,000 for the 100th-ranked occupation, and $17,000 for the 200th-ranked occupation.

The red line is the cumulative percentage of Navy officer personnel with respect to occupation rank. Reading up from the *x*-axis and over to the right-hand *y*-axis, we see that the top 400 occupations contain about 40 percent of the officers. The top 800 occupations have 67 percent of the officers, and the remaining 1,035 occupations have the remaining 33 percent.

We next consider whether high-ranked occupations in terms of average S&I pay also have a large share of S&I pay expenditure. This

Figure C.1
Distribution of Average S&I Pay and Distribution of Navy Officers for Four-Digit Occupations Ranked by Average S&I Pay (2012 Dollars)

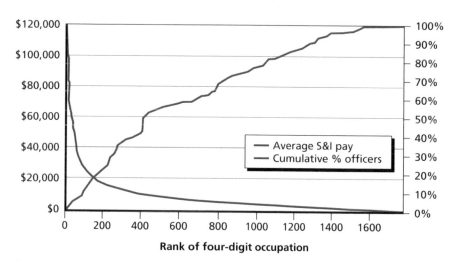

---

[3] There are three non–health care professions among the top 50, but these three represent only four officers.

need not be the case, but would be if occupations with high average S&I pay also tend to have a large number of officers. Figure C.2 shows the cumulative distribution of total S&I pay expenditure (the blue line), and for comparison it also includes the cumulative distribution of officers (the red line) shown in Figure C.1. Like Figure C.1, the ranking on the x-axis is based on average total S&I pay. As can be seen, the top 400 occupations have 40 percent of the Navy officers, and we now see that they account for 75 percent of total S&I pay expenditure. Thus, occupations with high average S&I pay also have a high share of total expenditure.

Table C.2 is a companion to Figure C.2 and presents information on NOBCs at the one-digit level or occupation field. The table shows the NOBC code range for each field, number of four-digit occupations in the field, number of individual officers that served in those occupations for 2002–2012, share of S&I expenditure going to the field, and share of officers in the field. Three fields account for 86 percent of S&I pay expenditure: health care services, aviation, and naval operations, the last of which includes nuclear-qualified officers. These are

Figure C.2
Distribution of S&I Pay Expenditure and Distribution of Navy Officers for Four-Digit Occupations Ranked by Average S&I Pay

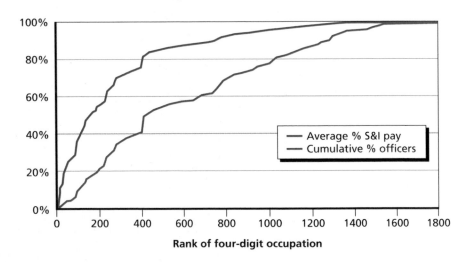

Table C.2
Distribution of S&I Pay Expenditure and Distribution of Navy Officers for One-Digit Occupations

Field Number	Occupation Field	NOBC Range	Number of Four-Digit Occupa-tions	Number of Officers	Share of S&I Expend-itures	Share of Officers
0	Health care services	1–999	201	25,188	38.1%	16.7%
1	Supply and fiscal	1000–1999	243	34,689	3.3%	13.2%
2	Sciences and services	2000–2999	166	11,568	3.2%	5.3%
3	Personnel	3000–3999	153	16,188	5.2%	7.4%
4	Facilities engineering	4000–4999	72	4,464	0.6%	1.8%
5	Electronics engineering	5000–5999	41	1,057	0.2%	0.4%
6	Weapons engineering	6000–6999	106	1,069	0.2%	0.5%
7	Naval engineering	7000–7999	81	3,225	1.3%	1.6%
8	Aviation	8000–8999	243	29,698	17.3%	17.3%
9	Naval operations	9000–9999	529	77,857	30.6%	35.9%
	Totals		1,835	205,003	100.0%	100.0%

large fields containing 70 percent of the officers and 973 of the 1,835 four-digit NOBCs. Table 6.1, presented in Chapter Six, is the same as Table C.2 but limited to the 400 top-ranked occupations. This subset of NOBCs accounts for 75 percent of total S&I expenditure (over all 1,835 occupations) and 40 percent of officers. Nearly all of this expenditure is in health care services, aviation, and naval operations; they account for 291 four-digit occupations, 70 percent of total expenditure, and 36 percent of officers. In sum, S&I expenditure is concentrated in the 400 top-ranked NOBCs and in three large fields within that 400.

Conclusion

S&I pay predominates in three occupational fields: health care services, aviation, and naval operations, the last of which includes nuclear-qualified officers. S&I pay in these fields is high on average and high as a fraction of total S&I expenditure. Health professions officers, aviation-related officers, and nuclear-qualified officers have sizable accession bonuses, incentive pay, and retention bonuses (see Chapter Two). If a wage differential initiative focused on these occupations, it would cover a significant fraction of officers receiving S&I pays. The incentive pay in effect operates as a wage differential, yet we also know that the retention bonuses make the cost-effectiveness of the total S&I pay greater than if it were all in a wage differential (see Chapter Four).

References

Asch, Beth J., James Hosek, and Craig Martin, *A Look at Cash Compensation for Active-Duty Military Personnel*, Santa Monica, Calif.: RAND Corporation, MR-1492-OSD, 2002. As of October 24, 2016:
http://www.rand.org/pubs/monograph_reports/MR1492.html

Asch, Beth J., Michael G. Mattock, and James Hosek, *The Blended Retirement System: Retention Effects and Continuation Pay Cost Estimates for the Armed Services*, Santa Monica, Calif.: RAND Corporation, RR-1887-OSD/USCG, 2017. As of January 22, 2018:
https://www.rand.org/pubs/research_reports/RR1887.html

Chetty, Raj, "A New Method of Estimating Risk Aversion," *American Economic Review*, Vol. 96, No. 5, December 2006, pp. 1821–1834. As of January 20, 2018:
https://www.aeaweb.org/articles?id=10.1257/aer.96.5.1821

Finkelstein, Amy, and Kathleen McGarry, "Multiple Dimensions of Private Information: Evidence from the Long-Term Care Insurance Market," *American Economic Review*, Vol. 96, No. 4, September 2006, pp. 938–958. As of October 24, 2016:
https://www.aeaweb.org/articles?id=10.1257/aer.96.4.938

Gould, Brandon, and Stanley Horowitz, "History of Combat Pay," in *Report of the Eleventh Quadrennial Review of Military Compensation: Supporting Research Papers*, Washington, D.C.: U.S. Department of Defense, 2012. As of August 6, 2018:
https://militarypay.defense.gov/Portals/3/Documents/Reports/11th_QRMC_Supporting_Research_Papers_(932pp)_Linked.pdf

Hosek, James, and Francisco Martorell, *How Have Deployments During the War on Terrorism Affected Reenlistment?* Santa Monica, Calif.: RAND Corporation, MG-873-OSD, 2009. As of August 7, 2018:
https://www.rand.org/pubs/monographs/MG873.html

Hosek, James, Shanthi Nataraj, Michael G. Mattock, and Beth J. Asch, *The Role of Special and Incentive Pays in Retaining Military Mental Health Care Providers*, Santa Monica, Calif.: RAND Corporation, RR-1425-OSD, 2017. As of September 22, 2018:
https://www.rand.org/pubs/research_reports/RR1425.html

Mattock, Michael G., James Hosek, Beth J. Asch, and Rita Karam, *Retaining U.S. Air Force Pilots When the Civilian Demand for Pilots Is Growing*, Santa Monica, Calif.: RAND Corporation, RR-1455-AF, 2016. As of September 11, 2018: https://www.rand.org/pubs/research_reports/RR1455.html

Rothschild, Michael, and Joseph Stiglitz, "Equilibrium in Competitive Insurance Markets: An Essay on the Economics of Imperfect Information," *Quarterly Journal of Economics*, Vol. 90, No. 4, November 1976, pp. 629–649. As of October 24, 2016:
http://www.uh.edu/~bsorense/Rothschild&Stiglitz.pdf

Simon, Curtis J., Shirley H. Liu, Saul Pleeter, and Stanley Horowitz, "Combat Risk and Pay: Theory and Some Evidence," in *Report of the Eleventh Quadrennial Review of Military Compensation: Supporting Research Papers*, Washington, D.C.: U.S. Department of Defense, 2012. As of August 6, 2018:
https://militarypay.defense.gov/Portals/3/Documents/Reports/11th_QRMC_Supporting_Research_Papers_(932pp)_Linked.pdf

Subramaniam, Anushree, *The Unaccounted Insurance Value of Medical Innovation*, Becker Friedman Institute for Research in Economics Working Paper Series No. 2016-04, Chicago: Becker Friedman Institute for Research in Economics, 2016. As of October 24, 2016:
https://papers.ssrn.com/sol3/papers.cfm?abstract_id=2771359

Under Secretary of Defense (Comptroller), *Financial Management Regulation*: Vol. 7A, *Military Pay Policy—Active Duty and Reserve Pay*, DoD 7000.14-R, Washington, D.C.: U.S. Department of Defense, 2017. As of October 16, 2018:
http://comptroller.defense.gov/Portals/45/documents/fmr/Volume_07a.pdf

U.S. Bureau of Naval Personnel, *Manual of Navy Officer Manpower and Personnel Classifications*: Vol. I, *Major Code Structures*, NAVPERS 158391, Washington, D.C.: U.S. Department of the Navy, January 2010. As of October 21, 2016:
https://navynavadmin.files.wordpress.com/2010/03/nocvol1.pdf

U.S. Department of Defense, *Report of the Tenth Quadrennial Review of Military Compensation*: Vol. 1, *Cash Compensation*, Washington, D.C.: U.S. Department of Defense, 2008. As of August 6, 2018:
https://www.hsdl.org/?view&did=713320

———, *Report of the Eleventh Quadrennial Review of Military Compensation: Main Report*, Washington, D.C.: U.S. Department of Defense, 2012. As of August 6, 2018:
https://militarypay.defense.gov/Portals/3/Documents/Reports/11th_QRMC_Main_Report_FINAL.pdf

———, *Aviation Incentive Pays and Bonus Program*, DoD Instruction 7730.67, Washington, D.C.: U.S. Department of Defense, 2016. As of January 20, 2018:
http://www.esd.whs.mil/Portals/54/Documents/DD/issuances/dodi/773067_dodi_2016.pdf

———, *Military Personnel Programs (M-1)*, Washington, D.C.: U.S. Department of Defense, 2017. As of August 6, 2018:
http://comptroller.defense.gov/Portals/45/Documents/defbudget/fy2017/marchAmendment/fy2017_m1a.pdf

U.S. Department of the Navy, "FY-16 Aviation Command Retention Bonus Program Information," memorandum, Washington, D.C.: U.S. Department of the Navy, September 16, 2015. As of May 19, 2016:
http://www.public.navy.mil/bupers-npc/officer/Detailing/aviation/OCM/Documents/FY16ACRBProgramInfo16Sep2015.pdf

U.S. Navy Personnel Command, "AvB ('The Bonus')," undated. As of January 20, 2018:
http://www.public.navy.mil/bupers-npc/officer/Detailing/aviation/OCM/Pages/ACCP.aspx

Varian, Hal R., *Microeconomic Analysis*, 3rd ed., New York: W. W. Norton & Company, 1992.